PLANET SOS

Marie G. Rohde

What on Earth Books

WoEB

What on Earth Books is an imprint of What on Earth Publishing.
The Black Barn, Wickhurst Farm, Tonbridge, Kent TN11 8PS, United Kingdom.
30 Ridge Road Unit B, Greenbelt, Maryland, 20770, United States.

First published in Spanish under the title: *Monstruos Verdaderos Que Amenazan El Planeta*
Text and illustrations copyright © Marie G. Rohde 2019
Layout copyright © 2019 Zahorí Books, Sicília, 358 1-A · 08025 Barcelona, Spain
www.zahorideideas.com

This edition first published in English by What on Earth Books in 2020.

Staff for this book: Editor: Katy Lennon; Designer: Joana Casals;
Cover Design: Andy Forshaw; Research consultant: Michelle Harris.

Library of Congress Cataloging-in-Publication Data available upon request

ISBN: 978-1-912920-22-8

Printed in Slovenia

10 9 8 7 6 5 4 3 2 1

whatonearthbooks.com

CONTENTS

SAVE OUR PLANET

Our planet is in grave danger. It is being threatened by monsters that are stomping around the Earth and causing complete chaos. These pesky beasts are busy at work creating storms, melting ice, drying out lakes, and filling the air with noise. And they feed off the strangest things: burps from cows, car exhaust fumes, plastic bags, garbage, and pollution. Humans make a lot of these things, and the more they make, the bigger and more confident the monsters become.

Recently, these planet-destroying beasts have been growing at an alarming rate. Early myths and legends tell tales of terrifying monsters that wreaked havoc on the planet, but humans were almost always there to rise to the challenge and defeat them. We can do it again! Read this book to arm yourself with knowledge. And as you do, keep an eye out for Monster Cards. They're your how-to guides for vanquishing the rampaging creatures that threaten our world.

It's time to take action and become an Earth-saving hero!

DON'T BE FOOLED . . .

Every monster has weaknesses. Check the Monster Cards for details of how to defeat them. All of the cards are at the back of the book. Use that section to plan your monster attack strategy.

MONSTER CARD

MONSTER MUG SHOT

MONSTER ICON (SEE MAP ON PAGES 48–49)

HABITAT
SKY
LAND
SEA

NAME — **OZONE SERPENT**

WHEN THE MONSTER WAS DISCOVERED — ·1985   [ozone depletion] — SCIENTIFIC NAME

DO THESE THINGS TO DEFEAT THE MONSTER (KEY ON PAGE 4)

THESE ARE THE THINGS THAT ARE FEEDING THE MONSTER (KEY ON PAGE 4)

MONSTEROMETER — 1 2 3 4 5 6

Don't forget to check out our modern monsters' mythological cousins in the mini-profiles throughout the book.

The Monsterometer shows you how much the monster is affecting our planet. If it has a reading of one, its actions are not causing us any harm. If it's showing six, however, then the monster is causing some really big problems and must be stopped.

MONSTEROMETER

I am the all-seeing, all-knowing Ozone Serpent. I glide around your planet nibbling on your delicious ozone layer. Some think I am a mysterious being as I am very good at hiding. One thing's for sure though—I am very powerful. I have been sucking away at the force-field of your ozone layer for years.

Your planet uses the ozone layer as a shield to protect it against the harmful rays of the Sun. Lucky for me, it is not very strong and I have managed to bite right through it. Now that your defenses are down, your planet is under attack by the Sun's rays.

Some years ago, you eagle-eyed humans spotted my gleaming teeth through the ozone hole. My cover was blown. You banned CFC (chlorofluorocarbon) gases that were used in things like refrigerators and spray cans. I used these to keep my teeth shiny and sharp—now I have toothache.

This ban, called the Montreal Protocol, is recognized by almost the whole world. But you still have a lot of work to do if you want to banish me for good. If the use of CFC gases doesn't stop completely, I'm sure I will be back to full strength in no time.

The Vikings believed that a dragon-like serpent called Niðhöggr [nith-hogg-er] chewed at the roots of Yggdrasil [ig-dra-sil], the world tree. If it ever finished, the tree would shiver and the end of times, Ragnarök [rag-ner-ak], would come.

ATMOSDRAGON

Planet Earth and I have always been great friends. We both hate the cold, so I wrap around the planet to keep it warm. In return I am given greenhouse gases that keep me healthy. Without me, Earth would be too cold for delicate creatures like humans to live on.

Recently, you humans have been taking really good care of me. You do lots of things that release greenhouse gases. Now I am full and happy. Driving cars and burning fossil fuels, such as coal and gas, make delicious carbon dioxide. Also, thanks to your taste for meat, I am treated to a particular delicacy—methane gas. This comes from cow and sheep burps and farts. Yum! However, the more gas there is for me to gobble up, the hotter Earth gets. This is called "global warming"—my favorite.

I have heard that some of you can't take the heat and want to put me on a diet. How rude! You say global warming is causing problems, such as ice caps melting, seas rising, and large storms sweeping over your homes.

I'm very worried. If you switch to renewable energy sources, like solar or wind power, I'll have less carbon dioxide to eat. And if you eat less meat, I won't have so much delicious methane. This will make me cool—and not in a good way.

Dragons from Chinese myths are powerful creatures that are seen as symbols of luck. They live in water and can control rain. It is thought that they can summon terrible storms.

ATMOSDRAGON

THE ACID SEA DRAGON

You will find me swimming silently through the ocean. I like the peaceful life, unlike my show-off twin, the Atmosdragon. He lives high in the sky, while I slink about under the sea. I also feed on the carbon dioxide that you humans make, and thankfully, there is lots in the ocean. The seawater absorbs gas from the air and I eat it up. I can then use it to make the water more acidic, which is the perfect habitat for me but is not so great for other sea creatures.

The Loch Ness Monster or "Nessie" is rumored to live in Loch Ness, a lake in Scotland. It is described as a shy creature, and only a few blurry photographs suggest its existence.

Over time, this acidic water slowly dissolves the skeletons and shells of creatures, such as clams, oysters, and coral. It can also stop their shells from forming in the first place. Humans seem to like these sea critters and are trying to stop me from harming them. They have made a pact to produce less carbon dioxide. All I want is to be left alone, but without carbon dioxide I will be very hungry, and annoyed . . .

ACID SEA DRAGON

THE LOGRE

When you hear the buzz of my saws and the rumble of my wheels, it's best to make yourself scarce. No tree or animal stands in my way, for I am the Logre.

Humans love plants and trees because they absorb carbon dioxide and make oxygen for people to breathe. But I hate them. They are just in my way. So are bumpy parts of the land. I uproot trees, flatten mountains, and alter the course of rivers. I have been working tirelessly for years, but I've not yet torn down half of all the trees on Earth. So much work left to do.

LOGRE

1 2 3 4 5 6

Humans are trying to stop me in my tracks, but they are too slow to catch me. Their efforts to protect forests are futile—I can swat them away like bugs.

Although I have heard on the tree vine that some know-it-alls are now recycling paper and replanting trees. Some people are even changing what they eat so that forests don't need to be cleared to raise livestock or grow crops. This is a bit of a worry for me—I'm going to have to start working longer hours!

The ogre from British fairy tale Jack and the Beanstalk is mostly interested in eating children and saying fee-fi-fo-fum. However, the nasty Logre prefers destroying forests and going brum-brrrrrum-brrrruuum.

THE ROAD SNAKE

ROAD SNAKE

I bet you use me almost every day, but have you ever really noticed me? Next time you're in your car, keep your eyes peeled. I'll be the one slithering beneath your wheels.

I've got the travel bug and my ambition is to visit every place in the world. I can stretch my asphalt for thousands of miles, crushing the earth and bringing the sweet sounds of traffic and the smell of pollution to any quiet, peaceful area. Cross me at your peril. I don't take kindly to trespassers.

Most people think I'm here to help, and I have become quite a celebrity in some parts. The homes of animals and plants are destroyed to make room for me. I delight when rainwater runs off my thick scales, taking pollution from the road into rivers and streams.

Continue using me for your gas-guzzling, fume-spewing vehicles, and we will go far. But stop using me and feel my wrath. If you decide to walk instead of drive, use electric cars, or campaign for more green spaces, I will be very, very disappointed in you.

Another famous serpent creature is Jörmungandr [your-mun-gan-dur]. This gargantuan, venomous snake from Viking legends was also intent on seeing the world. It was so large it could wrap its whole body around the Earth!

THE URBAN SPRAWLOSAURUS

I'm in the construction business—it's my job to find new areas of land to build on. I love sniffing out farmland that's just asking for some asphalt and concrete. My good friend the Road Snake helps me with this. Together, we are a formidable team.

In recent years, I have made great strides digging my claws into the earth around your cities. I've built lots of gorgeous suburban neighborhoods for you. So what's got into you lately? You've started grumbling about the very things you always loved. I thought we were friends. So what if building on farmland leaves less room to grow food? Or if all the grown-ups driving back and forth to work in the city make lots of greenhouse gases? My monster friends and I love that. Why can't you?

Sustainable city planning is something that I find quite scary. Pedestrian concourses, bike lanes, and improved public transportation are all things that could tame me. Some people are even adding green areas to their cities and limiting cars to reduce carbon dioxide. All of this is putting me out of work. Surely you'd like me to continue with my construction? Then we can be visited by new neighbors such as the Noisybird, Smogosaurus, or the Road Snake.

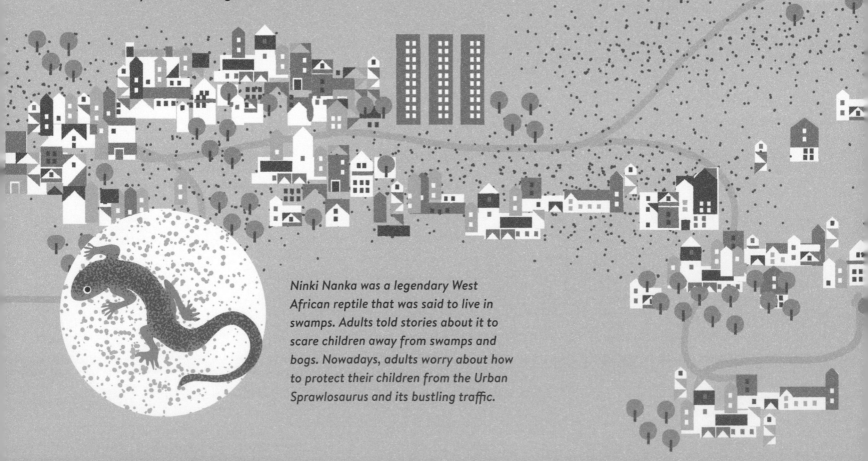

Ninki Nanka was a legendary West African reptile that was said to live in swamps. Adults told stories about it to scare children away from swamps and bogs. Nowadays, adults worry about how to protect their children from the Urban Sprawlosaurus and its bustling traffic.

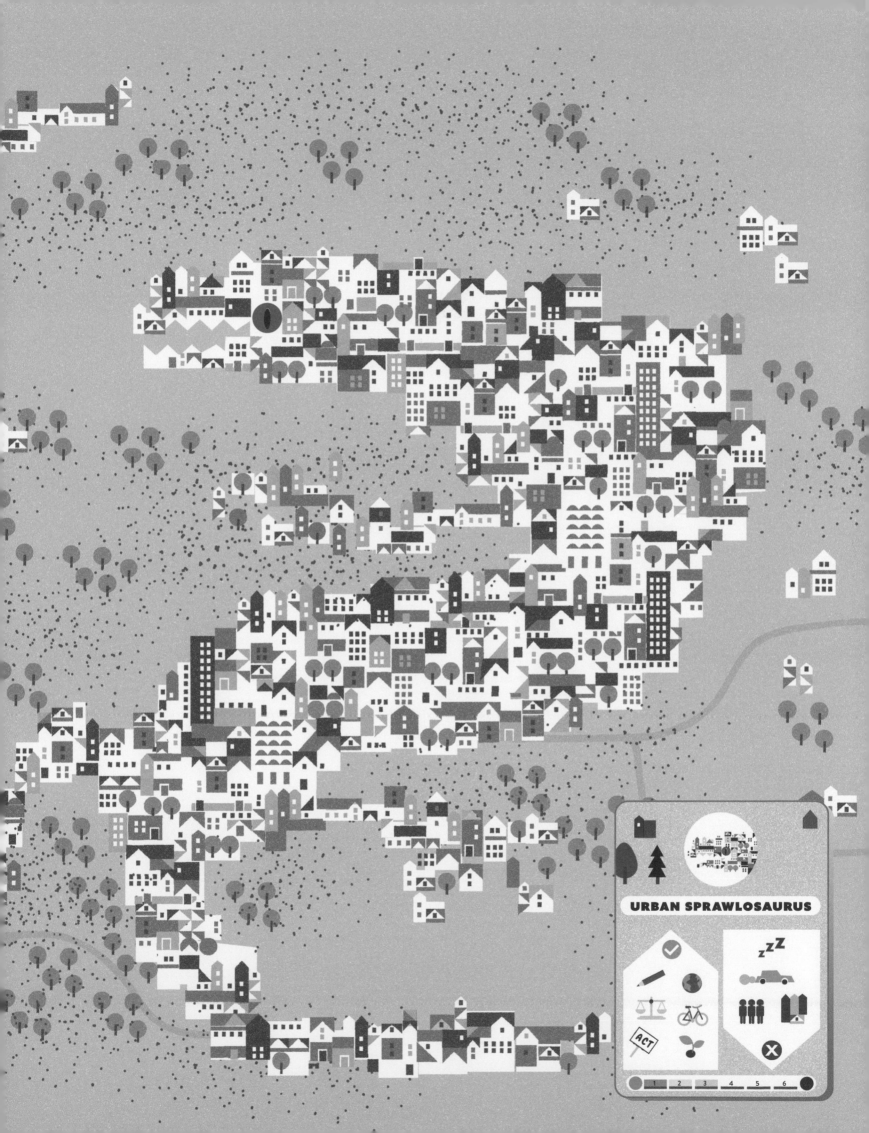

THE GLARE WORM

Let me introduce myself. I am the Glare Worm, a most brilliant beast. You'll struggle to find a star brighter than me, and just try to have a good night's sleep when I'm around. I'm the most bedazzling of monsters. This makes me beam with joy.

If you keep the lights on for me, I'll show off my sky-eclipsing powers. Artificial lights, such as street lights or lamps in your homes, make me glow even brighter. Please, dear humans, don't ever turn them off during the night—don't tell anyone, but I am afraid of the dark.

GLARE WORM

1 2 3 4 5 6

The night sky was swimming with stars before the invention of light bulbs, computers, televisions, and tablets. Your ancestors would use the sky to tell stories or navigate across the globe. Nowadays, if you want to gaze at the night sky or see our galaxy, the Milky Way, you'll have to travel far out into the countryside, or even to the ocean. In the cities you'll only find me, winking and blinking over your head. Help me stay radiant and shiny. Never turn off the light when leaving a room.

The Russian Firebird was known for the fiery glow of its feathers. If a feather was plucked from the bird, it would keep glowing and could light up a whole room.

THE NOISYBIRD

I honk, howl, roar, bang, shriek, pound, thump, and hammer every day in cities around the world. I can't stand silence—I like living my life loud and proud. I hate it when I can hear peaceful noises, such as birds singing. I need loud noise to survive, and I love the wail of saws, the honk of car horns, and the roar of airplanes.

It's my job to make sure noises are turned up to the max. If you want to chat to your friend, you'll have to shout. That's fine with me—the louder, the better. And there's no need for you to concentrate on your schoolwork or go to sleep. School and sleep are both overrated. When I visit the sea, I use sonar, or sound waves, to project my noise through the water. Troublemakers say this confuses sea creatures, causing them to move to different locations, and can even damage their hearing. But who needs sea creatures? They're not loud enough to be interesting.

So raise your voices and turn your speakers up loud to keep our cities and seas full of all those buzzing, crashing, whining, and other delightful noises. Keep your favorite monster (me!) alive and well.

In ancient China, the Jiu Tou Niao was a mythical bird with nine heads. It was nicknamed the "Ghost Carriage" because it made a loud noise like a vehicle at night-time.

SMOGOSAURUS

Good luck catching me—I'm made of smoke and dust. My body is built out of tiny particles that hover in clouds over your cities. I feed on exhaust fumes from cars and factories, and find smoke from burning coal extra tasty. I'm not fussy though. I'll also eat dust from construction work and suck up smoke from wildfires or volcanic eruptions. Children are not allowed to play outdoors when I lurk around with my toxic breath. I'll tickle your throat, making you cough. And I'll get in your eyes, making them sting. My party trick is to use my tail to block out the sun, making the blue sky seem gray. Impressive, huh? Sometimes when fog rolls in, I can make pollution so thick you can't even see through it.

I've been around for more than a century, but still humans haven't been able to snuff me out. Some are trying though. Many cities are limiting the numbers of trucks, cars, and other traffic that can drive through them. They are also improving public transportation and bike lanes so people can travel without causing air pollution. A campaign for renewable energy is even starting to gather force. If you want to say goodbye to me, that is a very good start.

The Smogosaurus's relative, the basilisk from medieval legends, was an extremely dangerous creature. Its toxic breath and stare would poison anything it came into contact with. Strangely, the only way to fight it was with a weasel—the basilisk hated the smell!

SMOGOSAURUS

ACID RAIN SPIRITS

Tlaloc [tla-lock] was the Aztec rain god, from the area that is now Mexico. A mask made of serpents hid his face. He could bring life and prosperity to Earth by sending rain, but he was feared as he could also bring destruction.

We used to be big celebrities—almost as famous as the Ozone Serpent. You could read about us in all the newspapers. We made forests wilt, destroyed buildings, and turned lakes so acidic that fish couldn't live there. And all we needed were a few gases, like sulfur dioxide and nitrogen oxide, which are released when humans burn fossil fuels. We could use our powers to turn the rain into some really wicked water.

Sadly, it seems as though we have had our time in the spotlight. Humans don't only get their energy from fossil fuels, they also use greener sources, such as wind or solar power. You've almost forgotten all about us. Sob! But there is always hope. Some parts of the world have not yet ended our reign. We still linger, drooling and waiting to make our comeback.

ACID RAIN SPIRITS

1 2 3 4 5 6

THE GREASE BEHEMOTH

Hi there! I'm your friendly neighborhood grease monster. I live below your feet in the sewers that run through the city. My steaming smell gives me away. Whoops, sorry about that! Some people say I smell a bit like a wet dog mixed with smelly feet. But I'm the perfect pet and very easy to feed: just flush snotty tissues, wet wipes covered in poo, bits of tangly hair, or cooking oil down your toilets

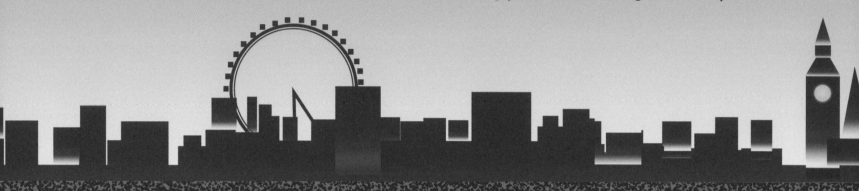

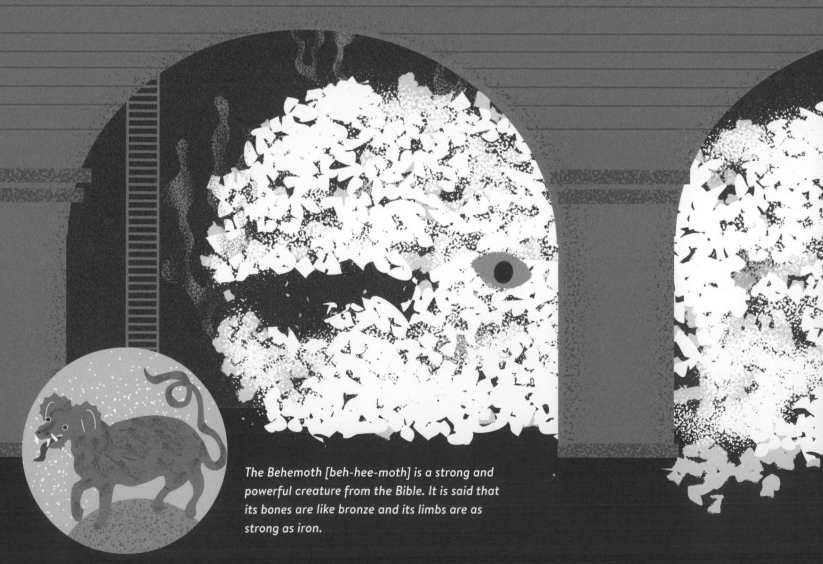

The Behemoth [beh-hee-moth] is a strong and powerful creature from the Bible. It is said that its bones are like bronze and its limbs are as strong as iron.

and sinks. These are treats to me. I will also feast on nail polish or coffee grounds if you let me.

Anything you send down here, I'll eat. But I'm a bit embarrassed—I have grown so massive that I've broken the pipes! Once in London, England, they caught me weighing nearly as much as a blue whale.

Sometimes you send down an army of nasty sewage workers to kick me out. The only things that will shift me are high-pressure hoses and shovels.

Will you come and visit me? In some cities, you can take guided tours in the sewers, and I've even been exhibited in some museums. You must like me after all!

GREASE BEHEMOTH

THE GRUBLIN

I am a lover of leftovers and a junk-food junkie. Any food that you don't eat goes straight into my tummy. Every year, around a third of all food that is produced for humans to eat goes into the garbage can. I never go hungry!

Growing crops and raising animals for food is a big business. It takes a lot of water, energy from fossil fuels, and land. If people didn't waste food, it would mean that more people could be fed with the same amount of work. But I don't like that.

My diet of cast-off food does make me quite gassy. As the food decomposes, it lets off the greenhouse gas methane. This gas can be captured and burned to create energy, but mostly it's just devoured by my friend the Atmosdragon. Humans are so picky. They won't eat any fruit or vegetable that is funny-looking. More for me, I don't care what they look like.

Composting food at home, instead of sending it to a landfill, is one way to stop so much methane filling the air. Also, using up leftovers and eating lopsided fruit and vegetables that would otherwise be thrown away will shrink me. Some say that I am a very greedy Grublin. And they are right. Keep the moldy munchies coming!

The Japanese nuppeppō [noo-pepp-ohh] is said to be a blob-like creature with terrible body odor. It only appears at night and enjoys shocking people with its pungent and nauseating smell!

TRASH KONG

Taller than a skyscraper and heavier than an airplane, I am the terrifying Trash Kong. Made from discarded garbage, I grow bigger by the second. You can feel the earth tremble and shake as I stomp through your cities.

Only one thing scares me, and that's the motto "reduce, reuse, recycle," which some of you humans have started chanting. This means that you are trying to buy fewer things, reuse and mend old things, and recycle anything that is waste. This leaves less trash, so I can't grow.

Thankfully, plenty of people are still careless with their waste, and masses of it is sent to landfill sites every year.

Many people only use things once before they throw them away. They also love being fashionable, so they throw away old clothes and buy new ones all the time; this is called "fast fashion." Nowhere is safe when I am around. I pollute water and soil as toxic chemicals leak out of my waste. Also, just like the Grublin, I give off large amounts of methane gas.

Lately, humans have really started to anger me. There has been talk of some countries banning single-use plastic objects. I've also heard people using ugly terms like "zero waste" and "second hand." You should know better than that. You should keep me happy. Or else . . .

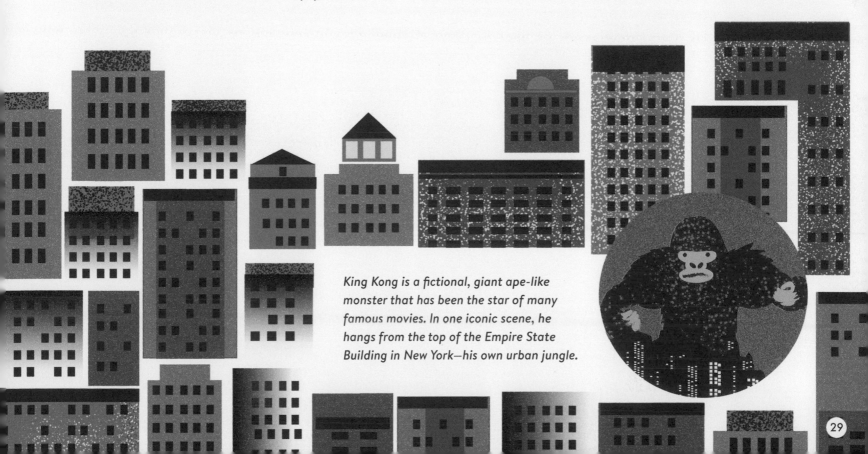

King Kong is a fictional, giant ape-like monster that has been the star of many famous movies. In one iconic scene, he hangs from the top of the Empire State Building in New York—his own urban jungle.

THE E-WASTE GOLEM

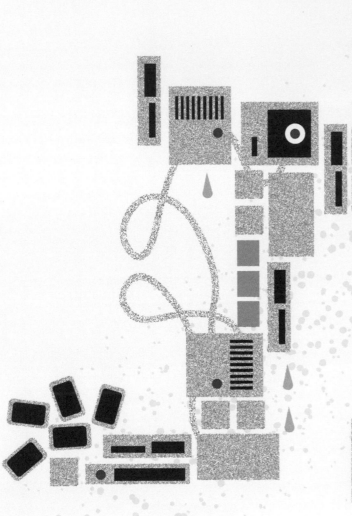

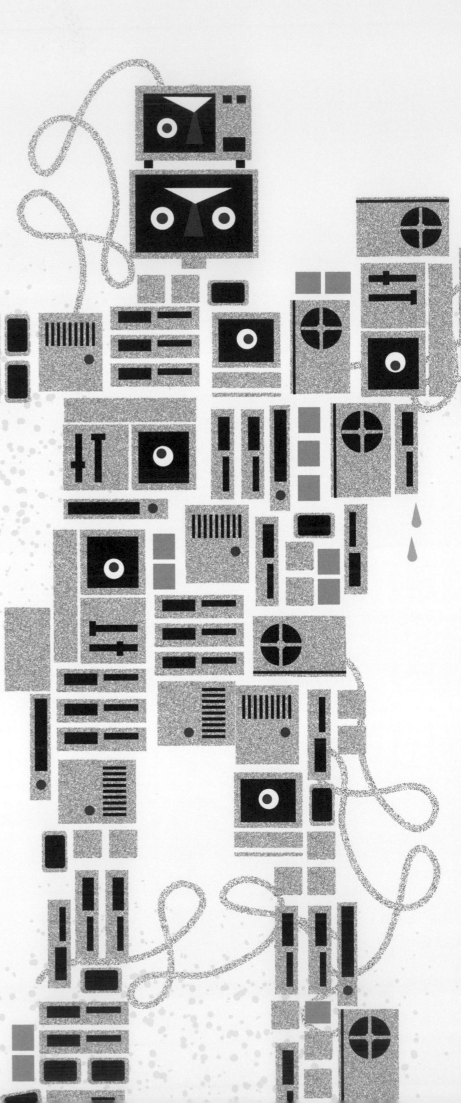

E-WASTE GOLEM

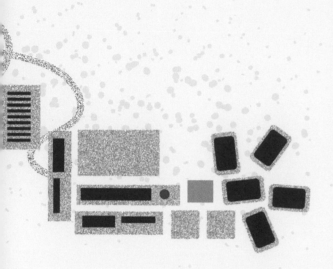

You threw me away but I'm back to haunt you. I am made of millions of electronic gadgets, batteries, chargers, mobile phones, computer monitors, and random electronic clutter. These things were expensive when you bought them, with valuable metals hidden inside.

You adored your gadgets and couldn't stop playing with them while they functioned. When they broke, grew old, or ran out of batteries, you threw them out and forgot about them. Now they're mine! You name it and I've put it to good use.

Sometimes I need to rest my weary wires, and when I do, I leak toxic metals, which pollute the air, water, and soil. You bury me in the ground so that you can't see me. But that doesn't mean I'm not here.

You humans could reuse my metals if you really wanted to. I could be a gold mine instead of trash. If you did that, then you wouldn't have to keep digging for metals and making new gadgets. Start recycling electronic waste—anything with a plug—and my motherboard will begin to malfunction. I cannot compute this madness!

Golems are creatures from Jewish folklore. They are servants that are made of clay and controlled by magic. They are simple beings and need exact orders to follow. In some stories, they turn on their masters and go on terrible rampages.

THE PLAKEN

Humans must really love me. They dump plastic waste and it flows into rivers around the globe, washing into the ocean and straight into my mouth. Some of my favorite dinners are plastic bags, water bottles, and plastic fibers from clothes. My tentacles are made of thousands of tons of plastic debris that I found floating in the oceans. I weigh nearly as much as 1,500 blue whales!

Even I am worried about how big I have grown. Soon there will be more plastic than fish in the ocean. Sea creatures and birds keep mistaking me for food and become trapped in my plastic tentacles.

I wouldn't mind if you stopped feeding me so much. Some people are trying. When fishermen catch pieces of me, they transport them to shore to be recycled. Many countries are even trying to ban single-use plastic items, like bags and straws. Some other people have invented floating vacuum cleaners to suck me out of the sea. Please help me with my diet by not littering, buying less plastic, and recycling as much as you can. Otherwise I will throw a tentacle tantrum!

According to Nordic lore, the Kraken was a huge octopus that trapped ships in its gigantic tentacles. The Plaken's tentacles are much worse. They reach everywhere and have even been found in some of the ocean's deepest trenches.

BLOOMING DOOM

Help! We algae are turning into a monster! We've always been so helpful to you. We float around in the water, minding our own business, and make oxygen for you to breathe. We soak up sunlight and feed on chemicals. We've never asked you for anything.

But now you people have been overfeeding us. You dump fertilizers, cleaning chemicals, and animal dung into rivers and lakes. And all your burning of fossil fuels gives us extra nitrogen

BLOOMING DOOM

1   2   3   4   5   6

to eat. Plus the Atmosdragon is warming the water, which makes us grow into huge crowded patches called blooms. There's not enough room for us all anymore.

Some of our blooms release poisons into the water, which can make humans and animals sick. Other blooms die and the process of their decay uses up oxygen, so other plants and creatures in the water can't breathe. We don't want this to happen. We are a peaceful bunch.

We'd be perfectly fine to go back to our old ways, making oxygen and being a happy part of the marine ecosystem. You can make that happen by encouraging your family to use eco-friendly cleaning products and picking up your pet's waste. Don't leave it on the street to be washed into the rivers. Thanks so much for helping us get back to the good old days!

The mythical Japanese kappa is a trickster who has an indent in its head that is full of water. The Kappa loves cucumber, so if you want to keep one happy, you just throw a cucumber into the water where it lives!

OIL SPILLATHAN

I am slick and smooth, black as the night, glossy and cool. You can find me seeping out from the Earth in areas where humans drill for oil. Humans are careless with me and often spill me when they try to move me from place to place.

I have been lying low in the mangrove forests of the Niger Delta, where the River Niger meets the Atlantic Ocean, for decades. No one can move me, no matter how hard they try. Once I decide to go somewhere, it is hard to stop me. At sea, I spread quickly on the surface of the water. I wrap around everything I touch, clinging like chewing gum. I am deadly to birds, mammals, and sea creatures.

Some humans have nicknamed oil "black gold," because it can be used to make many things and can be sold for lots of money. Oil is used as petrol to power cars, to heat homes, to make plastic for toys, and many more things. Humans are so dependent on it that it would take them a very long time to get used to a world without it. Of course, if they really wanted to, they could find alternatives. Bamboo toys are fine, right? And renewable energy works very well. But I think I'm safe for now. You would never change your life to get rid of me, would you?

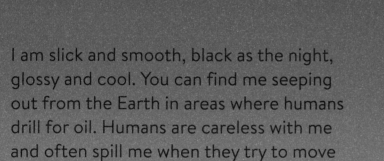

The Leviathan [leh-vai-ah-than] is a gigantic sea serpent from the Bible. It is able to cause chaos at sea, and in some stories it has more than one head.

THE AQUALIK

The tiddalik is a greedy, water-craving frog from
Australian Aboriginal legends. It gulps up all the
water around it, leaving a barren desert behind.
Stories tell of other animals that try to make it
laugh, so it will spit out the water for them to drink.

You live on a blue planet that is more than two-thirds covered with water. Most of it is either too salty to drink or frozen as ice. When humans waste or pollute precious fresh water, I appear. I'm all bad luck, evaporating and dirtying the water you need. MWAHAHAHAHA! The Atmosdragon raising the temperatures is also helping me by evaporating water.

I love it when you grow plants that need a lot of water, especially in dry places. You've helped me almost wipe out the Aral Sea by watering your crops. It was once Earth's fourth largest lake.

But now I'm scared. People are growing plants in their cities to catch water and help it soak into the soil. Plants also protect soil from the Sun, so water doesn't evaporate as quickly. Humans use plants to clean and reuse polluted water and collect rainwater to use for watering their crops. This gives me the creeps. Turning off faucets when brushing your teeth and other water-saving solutions also make me shiver. Perhaps I'm not as invincible as I think.

AQUALIK

THE DEGRADOTAUR

The soil on Earth used to be full of nutrients, but thanks to me, some of it is now barren and lifeless. It's my job to make sure that things can't grow. But there is still plenty of fertile soil left for me to ruin.

Cutting down trees to make room for animals or crops is a great idea of yours. Trees protect the land from erosion, and without them soil can be swept into rivers. Also, when there are too many farm animals eating the plants in an area, the soil suffers. Big tractors dragging their claws through the Earth is what I love to see, along with the ground being sprayed with pesticides and harsh chemicals. You humans don't bother trying to protect the soil for the future—I love it.

The Minotaur was a monster from Greek myths. It had the head and tail of a bull and the body of a man. It was so dangerous that it was imprisoned in a large maze by King Minos. It was defeated by the clever thinking of the king's daughter Ariadne and the strength of the hero Theseus.

Half of all the fertile topsoil has been lost in the last 150 years. And that's only the beginning. My evil plan is to make farming even harder in the future. Big farming companies that avoid being eco-friendly are my best friends forever.

One thing that is really getting in my way is sustainable farming. This is when people cut down on using toxic chemicals, give animals space to roam, and rotate the crops in a field. How am I supposed to do my job with this going on?

DEGRADOTAUR

THE OBLIVIONPEDE

I hate bugs—except myself. Little creepy crawlies give me the shivers. When you spray pesticides on your crops, bugs die off and I grow.

Bees are a particular annoyance for me. They pollinate a large number of the crops that humans eat, which helps the plants to grow. Also some birds and animals starve when you kill off the insects that they eat. But that's got nothing to do with me. Get rid of them, I say, so I can live a quiet, insect-free life.

OBLIVIONPEDE

1 2 3 4 5 6

In places where pesticides are used, you'll notice a strange quiet. You won't hear many birds singing or insects buzzing. A biologist called Rachel Carson found just this and published *Silent Spring* in 1962. This book was a wake-up call for people and led to a ban on certain pesticides. That was a hard blow to me. Since then there have been many new pesticide formulas, which give me hope. But each time there is a new effort to ban them. Some cities are even cutting back on pesticides, letting weeds sprout and insects and birds prosper. It's driving me crazy.

The Ōmukade [ohh-moo-kah-deh] was a mythical centipede from Japanese legend. It lived in the mountains and liked to eat humans!

NUCLEAR JINNS

We are the toxic and untamable Nuclear Jinns, and we are here to spread poison over your planet. Humans thought they were supersmart when they invented nuclear power. But then we appeared and caused them lots of trouble.

Nuclear power is a way of making energy by splitting atoms. It is a very efficient process and doesn't release greenhouse gases. Sounds great, huh? The downside for you humans is that it does create dangerous radioactive-waste products that are difficult to dispose of. That's us! We also harm any wildlife that we come into contact with. We Jinns are radioactive and deadly, and must be kept under strict control.

NUCLEAR JINNS

Humans have yet to eliminate nuclear waste and keep us Jinns at bay. Nuclear power plants need to be super secure to keep us inside, which means that they are expensive to build and run. And we are sneaky. We have already escaped a few times. Once free, we can fly for miles by hitchhiking on the wind. Everyone who lives in our path has to move away. We have even succeeded in making some areas completely uninhabitable.

One way to contain us is to bury us deep underground—but this doesn't always work. The biggest threat to Jinns would be if humans stopped making nuclear energy altogether and switched to renewable energy, like solar or wind power. But surely you don't want to do that. Do you?

The jinns from Arabic mythology were thought to be made of fire that didn't give off any smoke. They could be good or evil and, depending on their mood, they could grant you a wish or cause you harm.

AFTERWORD— CARBON BIGFOOT

You might have noticed that humans are the ones who have released all of the monsters in this book into the world. The impact that many of these monsters have on the Earth is called our carbon footprint. So that makes all of us Carbon Bigfoot monsters. Yikes!

But wait. We made those monsters, so we can vanquish them. Lots of

CARBON BIGFOOT

people are already working on it. Join them by reducing the amount that you buy or choosing things that are more eco-friendly. You can encourage the use of renewable energy, too. Try eating organic food or switching to a plant-based diet when you can. This will help save land so wild animals can thrive. There are also many other things that you can do. Just check the actions on the Monster Cards.

Our monster-shrinking quest starts with just a few small steps. But before you know it, those small steps will become giant leaps towards a healthier Earth. You have the power. It's time to act.

The Bigfoot of North American folklore is a tall, hairy creature with enormous feet. It is great at hiding and doesn't disturb the forests where it lives. We could all be more like Bigfoot and learn to protect the land we live on.

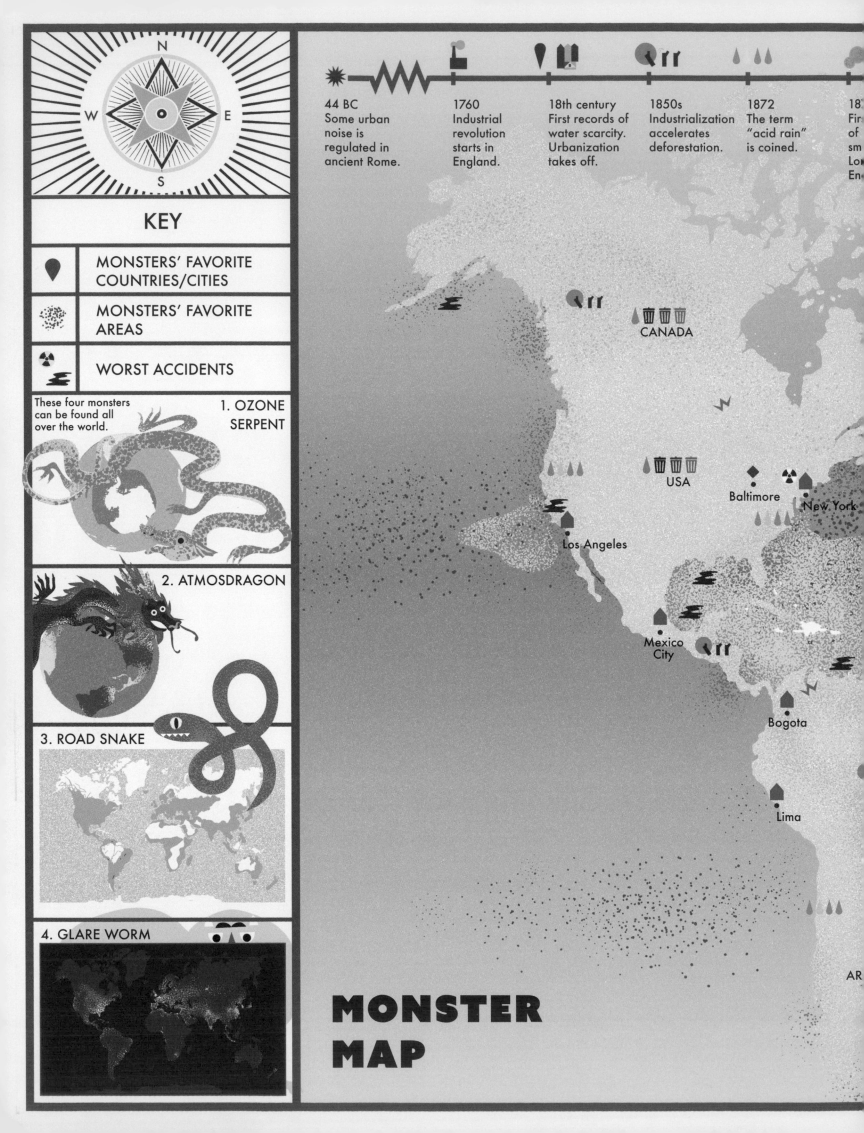

KEY

📍	MONSTERS' FAVORITE COUNTRIES/CITIES
⣿	MONSTERS' FAVORITE AREAS
☢	WORST ACCIDENTS

These four monsters can be found all over the world.

1. OZONE SERPENT

2. ATMOSDRAGON

3. ROAD SNAKE

4. GLARE WORM

44 BC
Some urban noise is regulated in ancient Rome.

1760
Industrial revolution starts in England.

18th century
First records of water scarcity. Urbanization takes off.

1850s
Industrialization accelerates deforestation.

1872
The term "acid rain" is coined.

18
Fir
of
sm
Lo
En

CANADA

USA

Baltimore

New York

Los Angeles

Mexico City

Bogota

Lima

AR

MONSTER MAP

GLOSSARY

CARBON DIOXIDE (CO_2)
Gas produced by chemical reactions and by humans and animals breathing out.

CFC GASES (chlorofluorocarbons)
Gases formerly used in refrigerators and spray cans. When released into the atmosphere, CFCs are broken up by the Sun's rays into atoms that destroy the ozone layer.

CLIMATE CHANGE
Process in which environmental conditions on Earth change over time.

COMPOSTING
Process of turning food waste, plants, or paper into compost—a nutrient-rich soil that is used to grow plants.

CROP ROTATION
Practice of changing the type of crop that is grown in an area to keep the soil healthy.

CROPS
Plants that are grown for humans to eat.

DECOMPOSE
Process in which something rots or decays.

DEFORESTATION
Cutting down trees in a large area, often to make space for growing crops or for livestock to graze.

ECO-FRIENDLY
Something or someone that is kind to the planet and doesn't harm the environment.

ECOLOGICAL FOOTPRINT
Impact of human activities on the environment.

ECOSYSTEM
Relationship between plants, animals, and their environment.

EMISSION
Production or release of something, usually a gas.

ENDANGERED
Animal or plant that is in danger of becoming extinct—dying out completely.

EROSION
Gradual destruction or removal of rock, ice, or soil by weather, such as wind or rain.

FERTILIZERS
Substances that help plants to grow. They can be chemicals or natural materials, such as horse manure.

FOSSIL FUELS
Non-renewable energy sources made from the decayed remains of plants or animals. Examples are oil and coal. They release large amounts of carbon dioxide when extracted or burned.

GLOBAL WARMING
Gradual rise in the temperature of Earth's surface due to high levels of greenhouse gases, such as carbon dioxide.

GREENHOUSE GASES
Gases such as methane or carbon dioxide that trap heat from the Sun. The heat in these gases makes the planet warm up at a higher rate, causing global warming.

HABITAT
Environment in which a plant grows or an animal lives.

INDUSTRIALIZATION
When lots of factories are built in an area.

INDUSTRIAL REVOLUTION
Period of time during the 18th century when many societies started building and using factories to make objects and food rather than making them by hand. Many rural areas were also urbanized, and railroads were built.

LANDFILL
Area where trash is taken and buried.

LIVESTOCK
Animals that are kept on a farm, for example, cows and sheep.

MANGROVE FORESTS
A group of trees and shrubs that grow along coastlines in slow-moving water.

MONTREAL PROTOCOL
An international agreement to protect the ozone layer by eliminating the use of harmful products, such as CFCs.

NUCLEAR
Something that uses the power created when the nucleus (central part) of an atom is split or combined.

OCEAN ACIDIFICATION
When the sea becomes acidic because it has absorbed carbon dioxide from the air. The more carbon dioxide in the air, the more acidic the ocean becomes.

ORGANIC
Something that has been farmed in a sustainable way or using natural products, rather than artificial chemicals, such as fertilizers.

PESTICIDES
Chemical substances used to protect plants from pests, like weeds, fungi, or insects.

POLLINATE
To fertilize a plant or tree by transferring pollen from one flower to another.

RADIOACTIVE
A substance that produces harmful energy.

RENEWABLE ENERGY
Energy created from sources that won't run out. This can be the wind, the Sun, falling water, waves, or heat from the Earth.

SECOND HAND
Something that has been previously owned or used by someone else.

SOIL DEGRADATION
Destruction of soil quality through non-sustainable farming, urbanization, or deforestation.

SUSTAINABLE
When natural resources are used at a steady level that doesn't damage the environment.

SYNTHETIC FABRIC
Fabric that is made of artificial fibers, such as plastic, rather than natural fibers, such as cotton.

TOPSOIL
Layer of soil that is closest to the surface.

URBANIZATION
When people come together to live in one place, forming cities.

SONAR
Equipment that is often used on ships to calculate the depth of the ocean or to find the position of something.

SUBURB
A residential area on the edge of a city or just outside it.

WATER SCARCITY
When there is not enough fresh water to meet human demand.

ZERO WASTE
The aim to use things that don't produce waste, or to recycle or reuse things so that nothing is sent to a landfill.

Boohoo!

I'm riding my bike to school. That means no greenhouse gas emissions.

I'm off to buy a sustainable snack. No plastic wrappers for you, Trash Kong!

GRRRR!

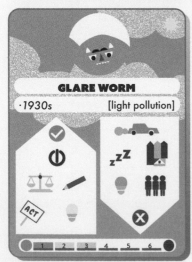

ACT NOW for a monster-free future. Help to clean up the planet by making a few small, everyday changes. Match the symbols on these cards to the key on the flap, so you know how to defeat each monster.

OZONE SERPENT

·1985 [ozone depletion]

1 2 3 4 5 6

ATMOSDRAGON

·1896 [global warming]

1 2 3 4 5 6

ACID SEA DRAGON

·1938 [ocean acidification]

1 2 3 4 5 6

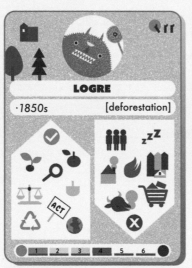

LOGRE

·1850s [deforestation]

1 2 3 4 5 6

ROAD SNAKE

·1950s [fragmentation]

1 2 3 4 5 6

URBAN SPRAWLOSAURUS

·18th century [urbanization]

1 2 3 4 5 6

GLARE WORM

·1930s [light pollution]

1 2 3 4 5 6

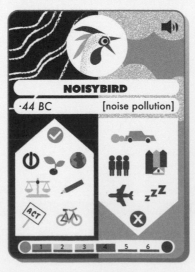

NOISYBIRD

·44 BC [noise pollution]

1 2 3 4 5 6

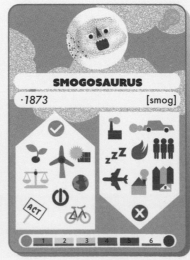

SMOGOSAURUS

·1873 [smog]

1 2 3 4 5 6

ACID RAIN SPIRITS

·1872 [acid rain]

1 2 3 4 5 6

I'm making my own sustainability plan. Some things I'll change right away, others might take more time.

We're lucky to live safe from harm in the Children's Eternal Rainforest reserve in Costa Rica. Children from 43 different countries have helped to save our home. The monsters didn't see that coming.

GREASE BEHEMOTH
2010s · [fatberg]

PLAKEN
·1938 · [plastic pollution]

DEGRADOTAUR
·1931 · [soil degradation]

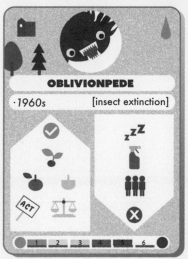

GRUBLIN
·1970s · [food waste]

BLOOMING DOOM
·c. 1870s · [harmful algal bloom]

OBLIVIONPEDE
·1960s · [insect extinction]

TRASH KONG
·1950s · [landfill]

OIL SPILLATHAN
·1910 · [oil spill]

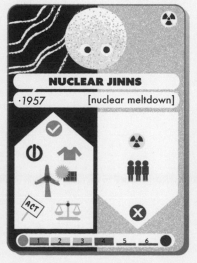

NUCLEAR JINNS
·1957 · [nuclear meltdown]

E-WASTE GOLEM
·1976 · [e-waste]

AQUALIK
·1800s · [water depletion]

CARBON BIGFOOT
·1990s · [ecological footprint]

INDEX

SOURCE LIST

NASA, Goddard Space Flight Center—Ozone Watch (https://ozonewatch.gsfc.nasa.gov/)

United Nations—Sustainable Development Goals (https://www.un.org/sustainabledevelopment/climate-change/)

US Environmental Protection Agency, Woods Hole Oceanographic Institution, 2016

Hansen, Potapov, Moore, Hancher, et al. University of Maryland, Department of Geographical Science—Global Forest Change (https://earthenginepartners.appspot.com/science-2013-global-forest)

The International Council for Science's Committee on Data for Science and Technology (https://sedac.ciesin.columbia.edu/data/collection/groads/maps/gallery/search)

United Nations (DESA/POPULATION DIVISION)—World Urbanization Prospects 2018 (https://population.un.org/wup/Maps/)

NASA Visible Earth maps, Night Lights 2012

Mimi Hearing Technologies, MIMI Worldwide Hearing Index 2017

World Health Organization—Global Ambient Air Pollution 2018 update (http://maps.who.int/airpollution/)

Likens, Butler, and Rury, Acid Rain, Encyclopedia of Global Studies (Ed. Anheier & Juergensmeyer 2012)

BCFN centre, The Economist, Intelligence Unit, Food Sustainability Index 2018 (Food waste at end-user level)

Kaza, Silpa et al, What A Waste 2. A Global Snapshot of Solid Waste Management to 2050, (2018) (Map of waste generation per capita. Greater than 1.5 kg.).

UNU, SCYCLE, ITU, ISWA, The Global E-Waste Monitor 2017 (Top ten producers of e-waste)

Dumpark, Sailing seas of plastic (https://app.dumpark.com/seas-of-plastic-2/)

Woods Hole Oceanographic Institution, Global distribution of PSP (Paralytic Shellfish Poisoning) toxins (2015)

Food and Agriculture Organization of the United Nations, Aquastat 2015

Food and Agriculture Organization of the United Nations, Nachtergaele, Petri, Biancalani, Van Lynden, and Van Velthuizen, (2010) Global Land Degradation Information System (GLADIS). Beta Version. An Information Database for Land Degradation Assessment at Global Level. Land Degradation Assessment in Dry lands Technical Report, No. 17

Food and Agriculture Organization of the United Nations, Pesticides Use (Top 10 Countries) 2016

Global Footprint Network, Ecological Footprint Per Person (https://www.footprintnetwork.org)

Prof. W. Kovarik, The Environmental History Timeline (http://environmentalhistory.org/)